D1299589

GRAPHIC DISCOVERIES

INCREDIBLE SPACE MISSIONS

by Gary Jeffrey

illustrated by Mike Lacey

rosen central™

The Rosen Publishing Group, Inc., New York

Published in 2008 by The Rosen Publishing Group, Inc.
29 East 21st Street, New York, NY 10010

Copyright © 2008 David West Books

First edition, 2008

Designed and produced by
David West Books

Editor: Gail Bushnell

Photo credits:
All photos courtesy of NASA

Library of Congress Cataloging-in-Publication Data

Jeffrey, Gary.
  Incredible space missions / by Gary Jeffrey ; illustrated by Mike Lacey. -- 1st ed.
      p. cm. -- (Graphic discoveries)
  Includes bibliographical references and index.
  ISBN-13: 978-1-4042-1090-5 (library binding)
  ISBN-13: 978-1-4042-9596-4 (6 pack)
  ISBN-13: 978-1-4042-9595-7 (pbk.)
  1. Astronautics--History--Juvenile literature. I. Lacey, Mike, ill. II. Title.
TL793.J45 2007
629.45--dc22
                                              2007004754

Manufactured in China

CONTENTS

SPACE RACE 4

EARLY SPACE TECHNOLOGY 6

THE FIRST SPACE WALK 8

APOLLO 11
THE MOON LANDING 15

APOLLO 13
DISASTER IN SPACE 29

SHUTTLES AND THE ISS 44

GLOSSARY 46

FOR MORE INFORMATION 47

INDEX and WEB SITES 48

SPACE RACE

After World War II, the United States and the Soviet Union took opposite sides in a war of spying and propaganda, called the "Cold War." Both sides recruited German scientists who had worked on the V-2 rocket program and hoped to conquer space to show their nation's superiority.

FIRST MAN IN SPACE

People in the U.S. believed that their technology was more advanced than that of the Soviets. It was, therefore, a great shock to the American public when, on October 4, 1957, the Soviets launched the first satellite into space. Four months later, after many failures, the U.S. finally launched a satellite, *Explorer I*, into space. The "Space Race" had begun. But on April 12, 1961, the U.S. suffered another dent in its national pride when a Soviet cosmonaut, Yuri Gagarin, became the first human in space. Alan Shepard, the first American astronaut, made it into space 23 days later. The Soviets continued to claim "firsts"–for the first woman in space (Valentina Tereshkova in 1963), and the first space walk (Aleksei Leonov in 1965).

MAN ON THE MOON

The U.S. started the *Apollo* program with a plan to beat the Soviets to a first manned moon landing. The Soviets landed unmanned probes on the moon before the Americans, but they had several failures and eventually dropped out of the race. On July 20, 1969, American astronaut Neil Armstrong stepped on the moon. America had at last beaten the Soviets in the final stage of the "Space Race."

1. The Soviet Sputnik, *the first satellite.*
2. *Yuri Gagarin and Valentina Tereshkova (inset).*
3. *Aleksei Leonov, making the first space walk.*
4. *The first American astronauts (Alan Shepard is top left).*
5. The Apollo 11 *mission landed on the moon. The astronauts were Neil Armstrong, Michael Collins, and Edwin Aldrin.*
6. *Edwin Aldrin on the moon, taken by Neil Armstrong.*

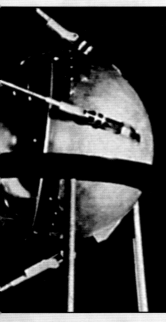

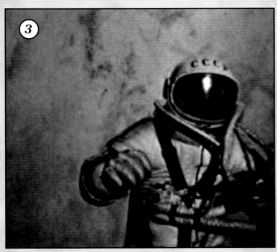

ПЕРВАЯ В МИРЕ
ЖЕНЩИНА-КОСМОНАВТ 1963
ВАЛЕНТИНА
ТЕРЕШКОВА

10к ПОЧТА СССР

EARLY SPACE TECHNOLOGY

Both the U.S. and the Soviets continued their space programs after the moon landing. Advances in technology saw a "car" driven on the moon, and people living in space stations which orbit the Earth.

COLD WAR THAWS IN SPACE

The U.S. sent six further missions to the moon, one of which didn't land (see page 29). The last was *Apollo 17,* sent in December, 1972. After this, manned space missions concentrated on orbiting space stations. Having decided against manned moon exploration, the Soviets had sent a series of *Salyut* space stations into orbit. Understanding the problems of living in space for long periods was essential for future long-distance, manned, space journeys. In 1973, the U.S. sent *Skylab,* their first space station, into orbit. Compared with the *Salyuts* it was massive, but from the beginning it was bugged with problems. It eventually fell to Earth in 1979. It was during the seventies that the two sides started to cooperate on joint space missions. In 1975, an *Apollo* and *Soyuz* docked together, and the crews shook hands. The Soviet Union went on to build the *Mir* space station. It was assembled in orbit between 1986 and 1996. Astronauts from many countries spent time on it until it finally fell to Earth in 2001.

Vostok 1
Yuri Gagarin made his historic flight in the Vostok 1 *spacecraft.*

Equipment module

Reentry capsule

Saturn V *launch vehicle (Apollo)*

Vostok 1 *launch vehicle*

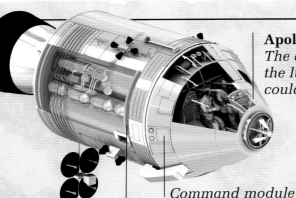

Apollo 11 *command/service module*
The command module docked with the lunar module so that the crews could transfer.

Ascent stage

Command module

Service module

Lunar rover
The Apollo 15 to 17 missions took a battery powered "car" to the moon. It was unfolded from the base of the descent stage and allowed the astronauts to explore much further.

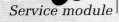

Descent stage

Lunar module
This was called the Eagle *on the Apollo 11 mission. When the astronauts left the moon the ascent stage split from the base (descent stage).*

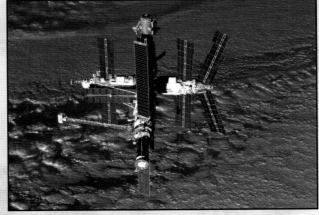

Orbiting space stations
Skylab *(left) had a set of solar panels ripped off at launch but still operated for six years.* Mir *(above) was very successful, despite having a spaceship crash into it and a fire on board!*

AS READY AS I'LL EVER BE, COLONEL BELYAYEV.

GOOD LUCK, ALEKSEI.

I'M CLOSING THE HATCH.

BEGINNING DEPRESSURIZATION.

...SHOULD TAKE ABOUT TEN MINUTES.

I WILL BE THE FIRST MAN TO WALK IN SPACE...

WHAT ARE THE DANGERS? I *KNOW* THE DANGERS.

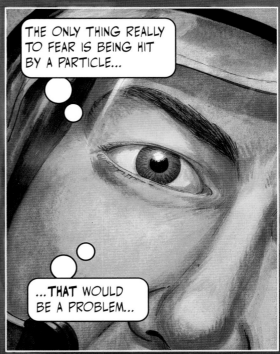

THE ONLY THING REALLY TO FEAR IS BEING HIT BY A PARTICLE...

...*THAT* WOULD BE A PROBLEM...

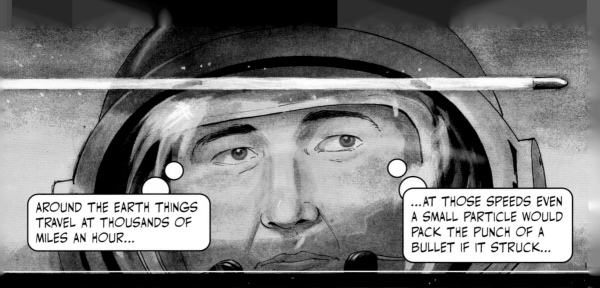

AROUND THE EARTH THINGS TRAVEL AT THOUSANDS OF MILES AN HOUR...

...AT THOSE SPEEDS EVEN A SMALL PARTICLE WOULD PACK THE PUNCH OF A BULLET IF IT STRUCK...

...GOING RIGHT THROUGH MY SUIT AND THEN THROUGH ME!

MY SUIT WOULD INSTANTLY DEPRESSURIZE...

HUFFFURK

...THE AIR WOULD BE SUCKED VIOLENTLY FROM MY LUNGS.

MY BODY WOULD SWELL AS THE FLUIDS IN MY ORGANS ESCAPE INTO THE VACUUM OF SPACE. IN 15 SECONDS I WOULD BE...

...AIRLOCK IS DEPRESSURIZED!

CCCP

FOR MOTHER RUSSIA...

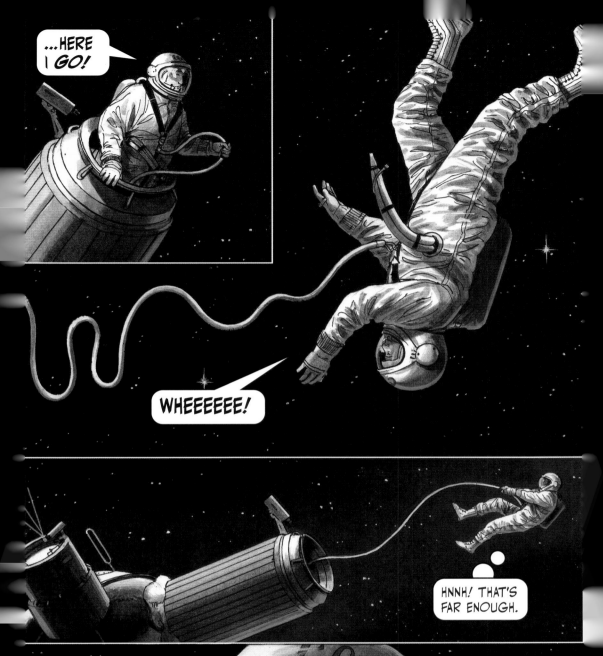

LEONOV, CAN YOU READ ME? WHAT'S HAPPENING?

I CAN'T BELIEVE THIS!

BELYAYEV, I'M STUCK. MY SUIT'S BLOWN UP SO MUCH I CAN'T BEND MY LEGS IN!

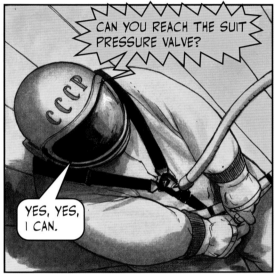

CAN YOU REACH THE SUIT PRESSURE VALVE?

YES, YES, I CAN.

BLEED OFF SOME AIR TO DEFLATE YOUR SUIT, BUT NOT TOO MUCH OR YOU'LL GET THE BENDS.*

*DECOMPRESSION SICKNESS.

OH, MY...

HISSSSSS!

NNNNGH!

UUGH! IT'S WORKED... I'M...... IN...

CLICK!

ALEKSEI, THE AIRLOCK HASN'T FULLY PRESSURIZED. IT MUST BE LEAKING!

NNNNGH...

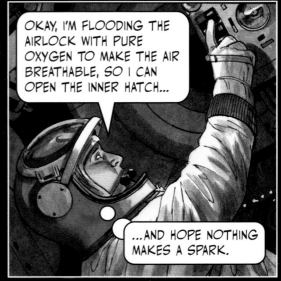

OKAY, I'M FLOODING THE AIRLOCK WITH PURE OXYGEN TO MAKE THE AIR BREATHABLE, SO I CAN OPEN THE INNER HATCH...

...AND HOPE NOTHING MAKES A SPARK.

AAGH, WELCOME BACK, COMRADE! FOR A MINUTE THERE...

WURRGH! PLEASE LET'S JUST JETTISON VOLGA...

...AND GET BACK TO EARTH.

AYE TO THAT!

THE END

APOLLO 11
THE MOON LANDING

MAY 25, 1961...

"I BELIEVE THIS NATION SHOULD COMMIT ITSELF TO ACHIEVING THE GOAL, BEFORE THIS DECADE IS OUT, OF LANDING A MAN ON THE MOON AND RETURNING HIM SAFELY TO EARTH."

JULY 16, 1969, LAUNCH COMPLEX 39-A, KENNEDY SPACE CENTER, FLORIDA...

PROJECT GEMINI

...MILLIONS OF DOLLARS, EIGHT YEARS OF RESEARCH AND IT BOILS DOWN TO THIS...

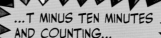

...T MINUS TEN MINUTES AND COUNTING...

EARLIER TODAY

...THREE MEN, PERCHED ON TOP OF A 3,000-TON ROCKET, AIMED AT THE MOON!

MISSION COMMANDER NEIL ARMSTRONG HAS AN ABORT HANDLE BY HIS KNEE, SO IF ANYTHING GOES WRONG DURING THE LAUNCH...

HEY, NEIL, WATCH YOUR POCKET ON THE HANDLE!

THANKS, MIKE. I'VE GOT IT.

MIKE COLLINS IS THE COMMAND MODULE PILOT.

YOU DON'T WANT THE MISSION TO END PREMATURELY!

HOLD ON TO YOUR HATS, WE'RE ABOUT READY TO SHAKE, RATTLE, AND ROLL!

BUZZ ALDRIN IS THE LUNAR MODULE PILOT.

09:32...

...10, ...9, ...IGNITING MAIN BOOSTER!

HOUSTON, WE HAVE NO COMPLAINT WITH ANY OF THE STAGES ON THAT RIDE...

...IT WAS BEAUTIFUL!

OKAY, TIME TO TURN THIS BABY AROUND.

APOLLO 11 IS TRAVELING AT 2,100 MILES (3,379 KILOMETERS) PER HOUR.

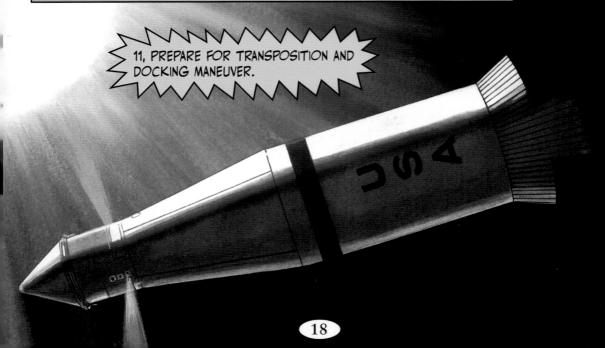

11, PREPARE FOR TRANSPOSITION AND DOCKING MANEUVER.

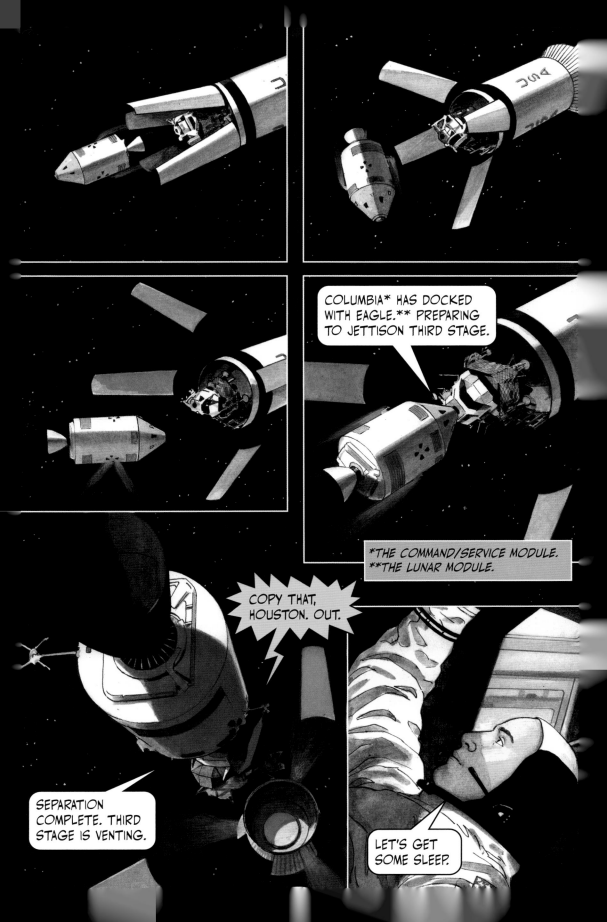

6,000 FEET (1,828 METERS) ABOVE THE LUNAR SURFACE...

HOUSTON, I HAVE A PROGRAM ALARM ...IT'S A 1202.

COPY THAT, NEIL.

STEVE?

IT'S JUST A COMPUTER OVERLOAD. TELL THEM TO KEEP GOING.

ASTRONAUT CHARLIE DUKE IS THE DUTY CAPCOM (CAPSULE COMMUNICATOR).

STEVE BALES IS A LUNAR MODULE FLIGHT CONTROLLER.

YOU ARE "GO" TO CONTINUE POWERED DESCENT.

JEEZ! THE SURFACE LOOKS SO CLOSE...

3,000 FEET (914 METERS) ABOVE THE LUNAR SURFACE...

ANOTHER ALARM...A 1201—CAN YOU GIVE ME A READING ON THAT?

21

OKAY, I'M GOING TO STEP OFF THE LEM* NOW.

THAT'S ONE SMALL STEP FOR...ZZZT...MAN...

*LUNAR EXCURSION MODULE.

...ONE GIANT LEAP FOR MANKIND.

NEIL ARMSTRONG'S MOON WALK IS WATCHED BY A WORLDWIDE TV AUDIENCE OF MORE THAN 600 MILLION.

15 MINUTES LATER...

BETTER GET TO WORK, BUZZ.

BEAUTIFUL, BEAUTIFUL. MAGNIFICENT DESOLATION.

24

"...AND NOW THE ASTRONAUTS PAUSE FROM THEIR EXPERIMENTS TO RAISE THE AMERICAN FLAG..."

I CAN'T GET THIS THING TO STAY STRAIGHT.

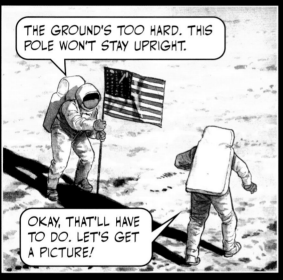

THE GROUND'S TOO HARD. THIS POLE WON'T STAY UPRIGHT.

OKAY, THAT'LL HAVE TO DO. LET'S GET A PICTURE!

PLEASE, *DON'T* FALL OVER!

21 HOURS AFTER LANDING...

11, WE ARE READY FOR LUNAR LIFTOFF IGNITION.

ERR, NEIL, I SEEM TO HAVE BROKEN THE FIRING SWITCH...

HERE, TRY JAMMING THIS PEN IN IT.

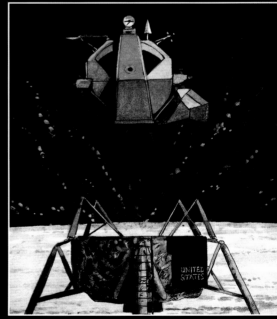

JULY 24...

SERVICE MODULE SEPARATION COMPLETE.

YOU ARE "GO" FOR REENTRY, 11.

RADIO BLACKOUT COMMENCES IN 3...2...1...ZZZT

THE COMMAND MODULE IS TRAVELING AT 25,000 MILES (40,000 KILOMETERS) PER HOUR.

"AS IT DESCENDS, THE OUTSIDE OF THE COMMAND MODULE WILL REACH MORE THAN 4,892°F. A SPECIAL HEAT SHIELD PREVENTS THE ASTRONAUTS FROM BEING BURNED ALIVE, ENABLING THEM TO..."

"...HOPEFULLY, SPLASH DOWN SOMEWHERE IN THE PACIFIC OCEAN."

27

13:45, THE NORTH PACIFIC.

THEY'RE ALIVE!

WELCOME HOME, FELLAS!

AFTER SPLASHDOWN, THE ASTRONAUTS SPENT A THREE WEEK PERIOD IN QUARANTINE TO MAKE SURE THEY WERE CLEAR OF ANY POSSIBLE DEADLY SPACE BUGS.

ISOLATION UNIT

HAILED AS HEROES, THE CREW WENT ON A 45-DAY TOUR OF MAJOR WORLD CITIES. PLANS WERE IN PLACE FOR A FURTHER SIX APOLLO MISSIONS TO THE MOON.

THE END

APOLLO 13
DISASTER IN SPACE

APRIL 13, 1970, 200,000 MILES (321,800 KILOMETERS) FROM EARTH, 55 HOURS 46 MINUTES INTO MISSION...

...I'D LIKE TO END OUR BROADCAST BY SAYING THAT, FAR FROM BEING A SCARY PLACE, OUTER SPACE CAN ACTUALLY BE A LOT OF FUN!

COMMANDER JAMES LOVELL

THIS IS THE CREW OF APOLLO 13, WISHING EVERYBODY AT HOME A NICE EVENING...

GENE KRANZ IS FLIGHT DIRECTOR OF WHITE TEAM.

ON THIS SPACECRAFT, WHAT DO WE HAVE THAT'S STILL GOOD?

THEY HAVE ENOUGH OXYGEN, FOOD, AND WATER TO LAST THE FOUR-DAY JOURNEY.

POWER IS GOING TO BE A PROBLEM.

THEY NEED TO MAKE SURE THERE'S ENOUGH CHARGE LEFT IN AQUARIUS'S BATTERIES* TO POWER ODYSSEY BACK UP WHEN THEY REACH EARTH, OR...

*UNLIKE ODYSSEY, AQUARUIS DIDN'T RUN OFF RENEWABLE FUEL CELLS.

...WE WON'T BE ABLE TO MAKE REENTRY —UNDERSTOOD!

HOUSTON, I WAS THINKING, WOULD IT BE POSSIBLE TO BURN THE ENGINE AFTER WE ROUND THE MOON, TO ER, YOU KNOW, SPEED US UP A LITTLE?

GOOD IDEA, JACK, WE'LL MAKE THE CALCULATIONS.

APRIL 14, 16:21...

ZZZT...THIS IS HOUSTON. LOSS OF SIGNAL BEGINS IN FIVE SECONDS...FOUR...

BYE, BYE, EARTH. SEE YOU ON THE OTHER SIDE.

...THREE...TWO...ZZZZZZZZZ ...SHHHHHHHHHHHHHHHHHH...

MOON'S SO CLOSE, YOU CAN ALMOST TOUCH IT!

FEAST YOUR EYES FRED, NO ONE'S COMING BACK UP HERE FOR A LONG TIME.*

*ACTUALLY APOLLO 14 DID, NINE MONTHS LATER.

8:23...

HOUSTON, WE'RE READY TO BEGIN OUR TRANSEARTH INJECTION SEQUENCE.*

*FIRING OF THE ENGINE.

...OKAY, WE JUST NEED YOU TO CHECK GIMBAL SETTINGS AGAINST YOUR A.O.T.*

OKAY, TAKING A LOOK NOW...

*ALIGNMENT OPTICAL TELESCOPE –FOR SIGHTING STARS.

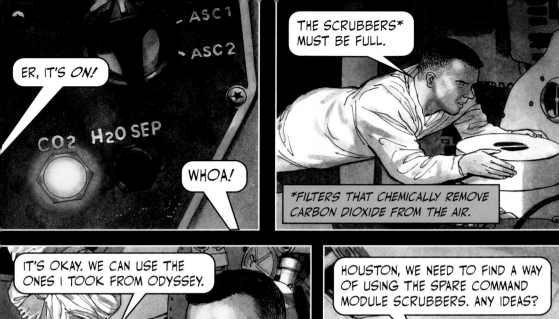

ER, IT'S ON!

CO2 H2O SEP

ASC 1
ASC 2

WHOA!

THE SCRUBBERS* MUST BE FULL.

*FILTERS THAT CHEMICALLY REMOVE CARBON DIOXIDE FROM THE AIR.

IT'S OKAY. WE CAN USE THE ONES I TOOK FROM ODYSSEY.

WE COULD, EXCEPT THEIR SOCKETS ARE THE WRONG SHAPE!

HOUSTON, WE NEED TO FIND A WAY OF USING THE SPARE COMMAND MODULE SCRUBBERS. ANY IDEAS?

WE'VE GOT SOMEONE WORKING ON IT. STAND BY.

WELL?

WE THINK IT'LL WORK...AS LONG AS THEY HAVE SOME DUCT TAPE ON BOARD.

AN HOUR LATER...

...ONE DO-IT-YOURSELF, LITHIUM HYDROXIDE UNIT, UP AND RUNNING!

APRIL 15, 12:35...

JODRELL BANK* SAYS APOLLO 13 IS COMING IN AT TOO SHALLOW AN ANGLE.

*AN EARTH TRACKING STATION.

ON HER PRESENT COURSE, SHE WILL MISS THE EDGE OF EARTH BY 99 MILES AND SKIM OFF INTO ENDLESS HIGH ORBIT...

13, WE NEED YOU TO MAKE A COURSE CORRECTION.

ROGER, HOUSTON. STARTING UP COMPUTER...

NEGATIVE! WE CAN'T SPARE POWER FOR THAT. YOU'RE GOING HAVE TO DO IT MANUALLY.

23:31...

OKAY 13, IGNITION...THRUST LOOKS GOOD...

15 SECONDS LATER...

OKAY, SHUT IT DOWN. NICE WORK.

LET'S HOPE IT WAS.

SHUTTLES AND THE ISS

As the cost of the space program increased, the U.S. looked for a cheaper alternative to getting people and satellites into space.

REUSABLE SPACECRAFT
A design for a spacecraft that could be reused was given the go-ahead in the 1970s. In 1981, the space shuttle *Columbia* flew into Earth's orbit and returned safely. A new era in manned spaceflight had begun, and a total of five space shuttles were built.

INTERNATIONAL SPACE STATION (ISS)
The Soviet Union collapsed in 1991, and the Cold War with it. A new spirit of cooperation in space meant that the U.S.A., Russia, and other countries around the world could now work together in building the ISS. The first part went into orbit in 1998, and it is due to be completed in 2010. Already, astronauts are spending six months at a time living and working in the ISS.

THE FUTURE IS *ORION*
After the loss of two shuttles, and because of the program's huge costs, the space shuttles will be taken out of service in 2010. A new spacecraft, called *Orion,* will replace them as part of Project Constellation, which will see us landing on the moon again.

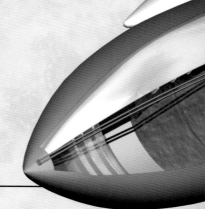

The ISS (above) is still being assembled in Earth's orbit. Replacement crews and supplies are ferried to the ISS by space shuttles (right), and by the Russian-built Soyuz TMA–7 *spacecraft (inset).*

Reusable
shuttle

Reusable solid fuel
rocket boosters

External fuel tank (not reusable)

Artist's impression of
Orion and the lunar
landing module.

GLOSSARY

abort To close down, or stop doing something, due to a problem.

airlock A small room with controllable air pressure and two entrances, which allows a person to pass between places with different air pressure, without air escaping.

altitude Height, usually above sea level.

amps Units of electrical current that produce a force between two parallel conductors placed three feet (1 meter) apart in a vacuum.

CO_2 Carbon dioxide. A gas that we breathe out, which is not good for our bodies.

confirm To agree, or state that something is correct.

conquer To gain possession over someone or something by force.

cosmonaut A Russian astronaut.

crater A hole left in the ground where a meteorite has landed.

cryo tank Extremely cold tanks, containing liquid oxygen under pressure.

debris What remains when something is crushed or destroyed.

docking Joining with another, such as one spacecraft to another.

epoch A period of time.

fuel cell A device that produces an electric current from a chemical reaction, using oxygen and hydrogen.

gravity The force that exists between two bodies.

ignition Starting an engine by lighting it, usually with a spark.

jettison To throw cargo overboard in order to lighten a vehicle.

particle A tiny amount of matter, such as a molecule, atom, or electron.

pressurized Compressed to a certain amount to maintain, in this case, an atmosphere that humans can survive in. When entering space from an airlock, the air has to be let out slowly. This is called depressurization.

probe An unmanned spacecraft that records and sends to Earth information about the environment it is passing through.

quarantine A place where people are kept away from others, so that germs cannot be passed on.

renewable Something that can replenish or replace itself.

satellite A man-made device that is launched by a rocket into space. It is placed in orbit around a planet for communication purposes.

transposition When two things are switched around.

vacuum A space where there is nothing at all.

velocity Speed.

venting Discharging or blowing out something.

FOR MORE INFORMATION

ORGANIZATIONS

Kennedy Space Center Visitor Complex
SR 405
Kennedy Space Center
FL 32889
(321) 449-4444
Web site: http://www.kennedyspacecenter.com

Smithsonian National Air and Space Museum
National Mall Building
Independence Avenue at 6th Street, SW
Washington, D.C. 20560
(202) 633-1000
E-mail: info@si.edu
Web site: http://www.si.edu/visit

FURTHER READING

Davis, Amanda. *Spaceships*. New York: PowerKids Press, 1997.

Engelhardt, Wolfgang. *The International Space Station: A Journey into Space*. Quadrillion Media LLC, 1998.

Hibbert, Clare. *The Inside and Out Guide to Spacecraft*. Chicago: Heinemann-Raintree, 2006.

Kuhn, Betsy. *The Race for Space: The United States and the Soviet Union Compete for the New Frontier*. Minneapolis: Twenty-First Century Books (CT), 2006.

Parker, Steve. *20th Century Science and Technology: 1960s Space and Time*. Milwaukee: Gareth Stevens Inc, 2001.

INDEX

A

Aldrin, Edwin, "Buzz," 4, 16, 24
Apollo missions, 28, 43
Apollo program, 4
Apollo 11, 4, 7, 15–28
Apollo 13, 29–43
Apollo 14, 36
Apollo 15 to 17, 7
Apollo 17, 6
Armstrong, Neil, 4, 16–28

B

Bales, Steve, 21
Belyayev, Colonel, Pavel, 9–14

C

Cold War, 4, 44
Collins, Michael, 4, 16
Columbia, 19, 44
Command module, 7, 16, 19, 27, 30, 31, 33, 39

D

Duke, Charlie, 21

E

Eagle, 7, 19, 20, 23, 26
Explorer I, 4

G

Gagarin, Yuri, 4, 6

H

Haise, Fred, 31, 33, 36, 37
heat shield, 27, 42

I

Iwo Jima, 43

J

Jodrell Bank, 40

K

Kennedy, President, John, 15
Kennedy Space Center, 15
Kerwin, Joe, 30
Kranz, Gene, 35, 38

L

Leonov, Alexei, 4, 8–14
Lousma, Jack, 30, 34, 35
Lovell, James, 29, 30, 38
lunar excursion module, 24
lunar landing module, 45
lunar module, 7, 16, 19, 30, 31, 33, 34, 38
lunar module lifeboat, 33
lunar rover, 7
Lunney, Glynn, 33

M

Mir space station, 6, 7
Moon landing, 4, 6
Moon walk, 24–25

O

Odyssey, 30, 32–35, 39
Orion, 44, 45

P

Project Constellation, 44
Project Gemini, 15

R

re-entry, 27, 35

S

Salyut space stations, 6
service module, 7, 19, 27, 41
Shepard, Alan, 4
Skylab, 6, 7
Soyuz, 6, 44
Space Race, 4
space shuttle, 44, 45
space walk, 4
Sputnik, 4
Swigert, John, 31

T

Tereshkova, Valentina, 4
Tranquility base, 23

V

Volga, 8, 14
Voskhod 2, 8
Vostock 1, 6
V-2 rocket, 4

Web Sites

Due to the changing nature of Internet links, the Rosen Publishing Group, Inc., has developed an online list of Web sites related to the subject of this book. This site is updated regularly. Please use this link to access the list:
http://www.rosenlinks.com/gd/space